我与家人朋友

罗国庆 陈良萍 编著

山东大学出版社

图书在版编目（CIP）数据

我与家人朋友 / 罗国庆，陈良萍编著 .
—济南：山东大学出版社，2014.10
（宝宝嘻哈乐学丛书）
ISBN 978-7-5607-5070-5

Ⅰ. ①我…
Ⅱ. ①罗… ②陈…
Ⅲ. ①家庭生活—儿童读物
Ⅳ. ① TS976-49

中国版本图书馆 CIP 数据核字（2014）第 149803 号

策划编辑：刘森文
责任编辑：刘森文 郑琳琳
封面设计：祝阿工作室

出版发行：山东大学出版社
社 址：山东省济南市山大南路 20 号
邮 编：250100
电 话：市场部（0531）88364466
经　　销：山东省新华书店经销
印　　刷：济南新先锋彩印有限公司
规　　格：880 毫米 ×1230 毫米　1/16
4 印张　　53 千字
版　　次：2014 年 10 月第 1 版
印　　次：2014 年 10 月第 1 次印刷
定　　价：18.00 元

版权所有，盗印必究
凡购本书，如有缺页、倒页、脱页，由本社营销部负责调换

目录

出入迷宫 — 找妈妈

帮助孩子找到妈妈。画出从起点到终点的路径．（Help the child find his mom. Draw a line from the start to the end.）

家庭关系一回答问题

看下面的图片，然后用“是”或“不是”回答问题。（Look at the picture and complete the following puzzle by writing “yes” or “no”.）

老张　老黄

小李　小张

小花　小宝　小草

1. 小宝的爷爷很年轻。（Xiao Bao’s grandfather is young.）

2. 小花的姐姐是短头发。（Xiao Hua’s sister has short hair.）

3. 老张和老黄有一个孩子。（Lao Zhang and Lao Huang have one child.）

4. 小李的爸爸是小花的奶奶。（Xiao Li’s father is Xiao Hua’s grandmother.）

5. 小花的爸爸是小李的兄弟。（Xiao Hua’s father is Xiao Li’s brother.）

6. 小花的爸爸的妈妈有长头发。（Xiao Hua’s father’s mother has long hair.）

7. 小草的兄弟的爸爸是长头发。（Xiao Cao’s brother’s father has long hair.）

8. 小花的兄弟的妈妈的孩子是年轻人。（Xiao Hua’s brother’s mother’s children are young.）

9. 小花的兄弟是白头发。（Xiao Hua’s brother has white hair.）

10. 小花是小宝的妹妹。（Xiao Hua is Xiao Bao’s sister.）

11. 小花有三个兄弟。（Xiao Hua has three brothers.）

12. 小宝的妈妈是短头发。（Xiao Bao’s mother has short hair.）

13. 小张和小李有三个孩子。（Xiao Zhang and Xiao Li have three children.）

14. 小草的奶奶是黑头发。（Xiao Cao’s grandmother has black hair.）

15. 小张的妈妈是小宝的奶奶。（Xiao Zhang’s mother is Xiao Bao’s grandmother.）

找出不同 — 爸爸妈妈和孩子

找出下面两幅图片的六个不同之处。（Find the six differences in the two pictures below.）

模仿涂色—驾驶汽车

按照第一幅图片的颜色，给第二幅图片中空白的汽车和男孩涂色。(According to the first picture above, color the blank car and boy in the second picture.)

自制小书 — 我是怎么出游的

在空白横线上填写如下词语：走路、公交车、小汽车、船、自行车、火车、飞机，然后按照背面介绍的方法制作一本关于旅行方式的迷你书，并读一读。（Fill in the words “walk, bus, car, boat, bike, train, airplane” in the blank. Then make a mini-book about how to get to school according to the method on the next page. Read it.）

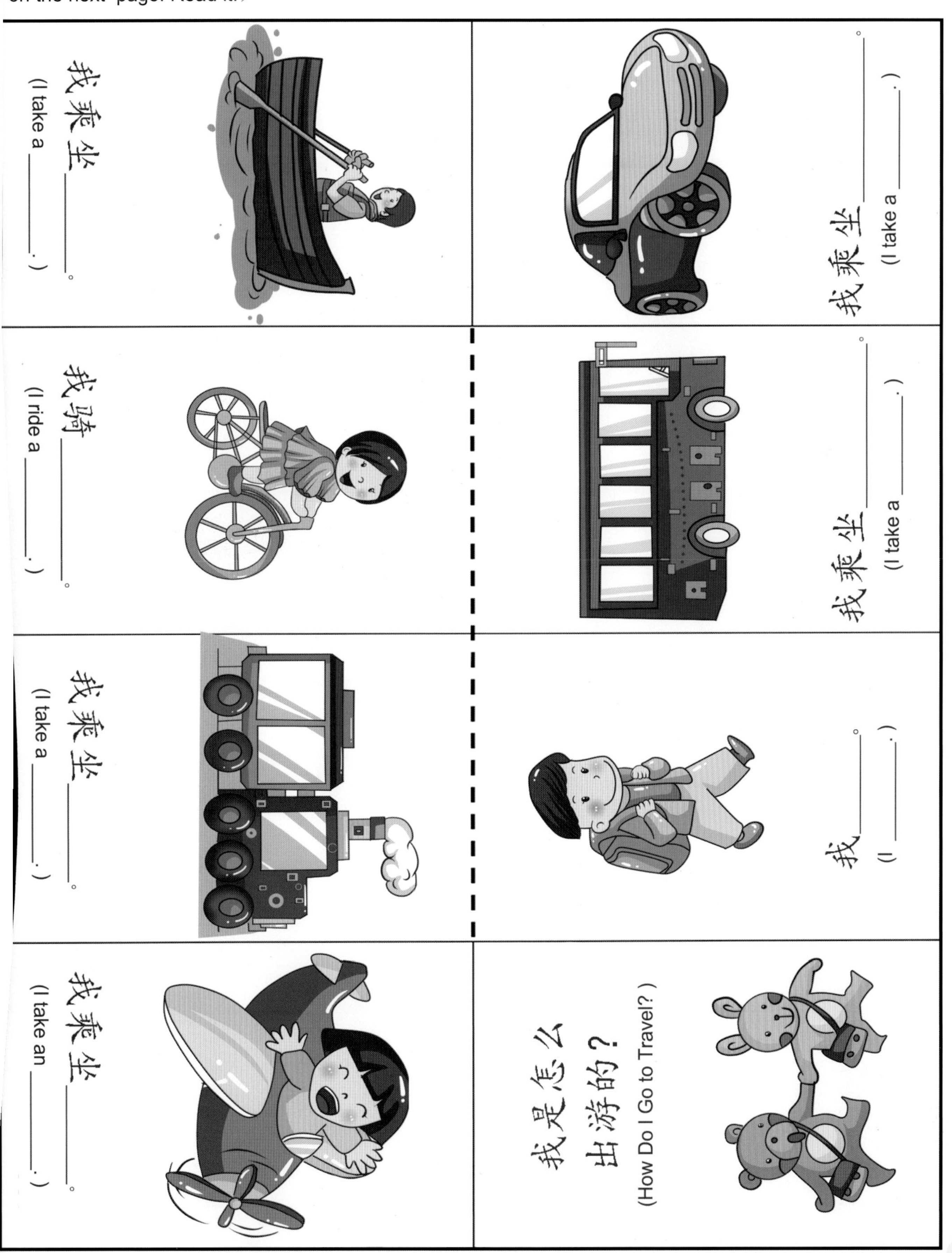

制作迷你书的方法
(How to Make Your Mini-Book)

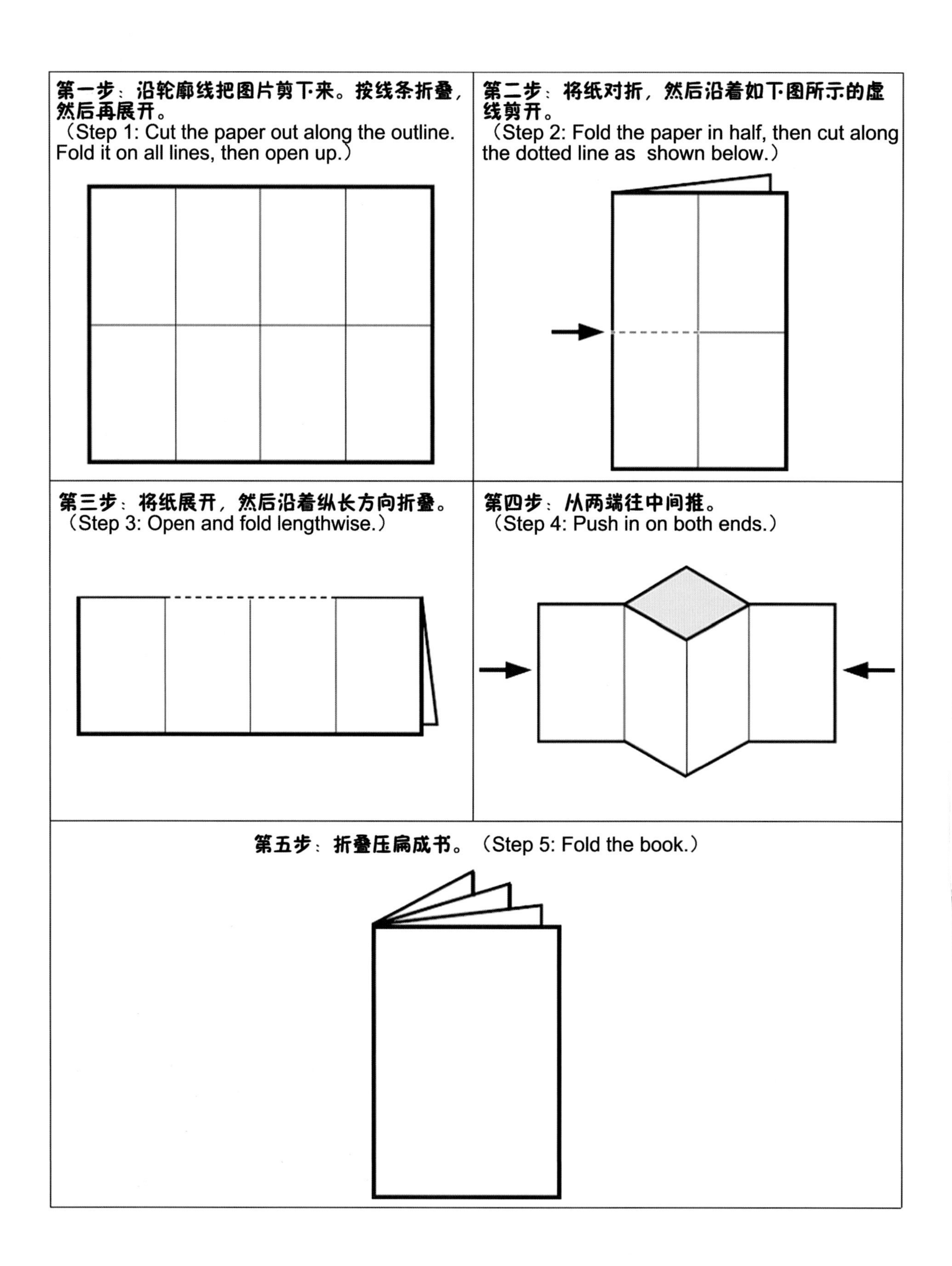

推断物品—小女孩的宠物

请读下面的小故事，然后从图片中圈选出正确的物体。（Please read the short story below and then circle the correct object from the pictures.）

一个小女孩问她父母，她是否能有一个宠物。爸爸妈妈同意她可以有一个宠物，只要宠物不是太大，能装在她的房间的小水池中就行。哪一个动物是小女孩的最佳宠物呢？（A little girl asks her parents if she could get a pet. Her dad and mom agree she can have a pet as long as it isn’t too big and can fit in a small tank in her room. Which animal will make the best pet for the little girl?）

出入迷宫一去幼儿园的路

帮助小男孩找到去幼儿园的路。（Help the boy find his way to the kindergarten.）

匹配连线 — 学习用品

将单词与对应的图片用直线连接起来。（Match the words to their pictures through a line.）

尺子 (ruler)

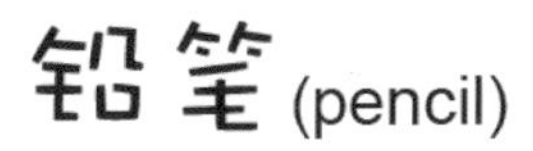

铅笔 (pencil)

订书机 (stapler)

剪刀 (scissors)

蜡笔 (crayon)

时钟 (clock)

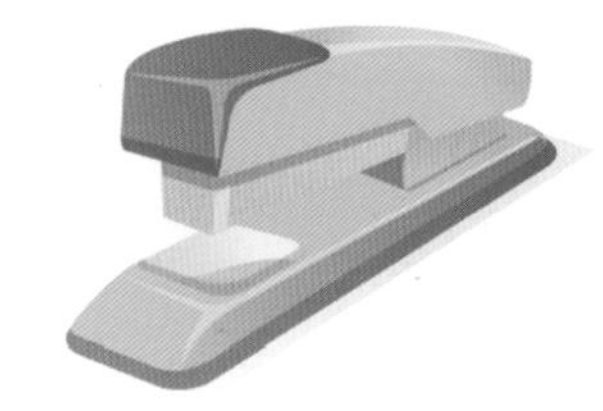

书包 (schoolbag)

纸 (paper)

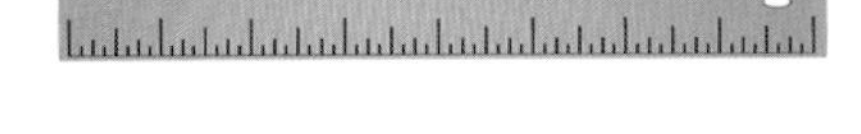

胶水 (glue)

胶带 (tape)

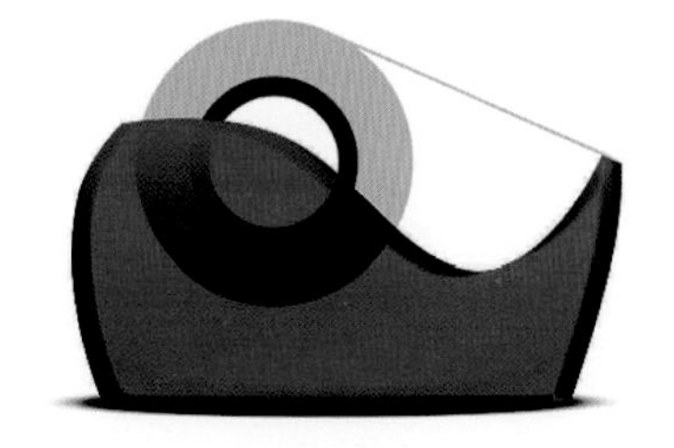

书 (book)

桌子 (desk)

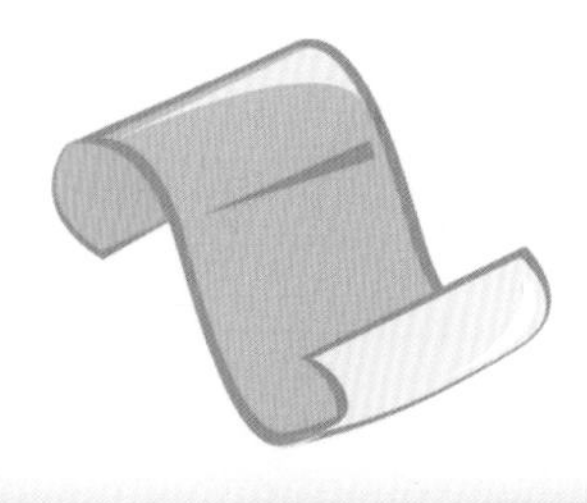

看图故事—你看到了什么

宗长给孩子讲下面的故事，同时给孩子指出下页图中的各种动物。（Tell the following story to the child, and point out the relative animals on the next page.）

棕熊，棕熊，你看到了什么？（Brown bear, Brown bear, what do you see? ）
我看见一只红色的鸟儿在看着我。（I see a red bird looking at me.）

红色的鸟儿，红色的鸟儿，你看到了什么？（Red bird, red bird, what do you see? ）
我看见一只黄色的鸭子在看着我。（I see a yellow duck looking at me. ）

黄色的鸭子，黄色的鸭子，你看到了什么？（Yellow duck, yellow duck, what do you see? ）
我看见一匹蓝色的马儿在看着我。（I see a blue horse looking at me. ）

蓝色的马儿，蓝色的马儿，你看到了什么？（Blue horse, blue horse, what do you see? ）
我看见一只绿色的青蛙在看着我。（I see a green frog looking at me. ）

绿色的青蛙，绿色的青蛙，你看到了什么？（Green frog, green frog, what do you see? ）
我看见一只紫色的猫在看着我。（I see a purple cat looking at me. ）

紫色的猫，紫色的猫，你看到了什么？（Purple cat, purple cat, what do you see? ）
我看见一只黑色的狗在看着我。（I see a black dog looking at me. ）

黑色的狗，黑色的狗，你看到了什么？（Black dog, black dog, what do you see? ）
我看见一只白色的绵羊在看着我。（I see a white sheep looking at me. ）

白色的绵羊，白色的绵羊，你看到了什么？（White sheep, white sheep, what do you see? ）
我看见一条金鱼在看着我。（I see a goldfish looking at me. ）

金鱼，金鱼，你看到了什么？（Goldfish, goldfish, what do you see? ）
我看见一只猴子在看着我。（I see a monkey looking at me. ）

猴子，猴子，你看到了什么？（Monkey, monkey, what do you see? ）
我看见孩子们在看着我。（I see children looking at me. ）

孩子们，孩子们，你们看到了什么？（Children, children, what do you see? ）
我们看见了一只棕熊、一只红色的鸟儿、一只黄色的鸭子、一匹蓝色的马儿、一只绿色的青蛙、一只紫色的猫、一只黑色的狗、一只白色的绵羊、一条金鱼、一只猴子在看着我们。（We see a brown bear, a red bird, a yellow duck, a blue horse, a green frog, a purple cat, a black dog, a white sheep, a goldfish, and a monkey looking at us. ）

这就是我们所看到的。（That's what we see.）

酒店

请按照上页的颜色，给本页空白的图片涂色。（According to the colorful picture on the previous page, color this blank picture.）

看图猜谜——生活谜语

根据下面带有插图的答案来猜谜语。（Find the answers to the riddles.）

我是书。
(I'm a book.)

我是毛衣。
(I'm a sweater.)

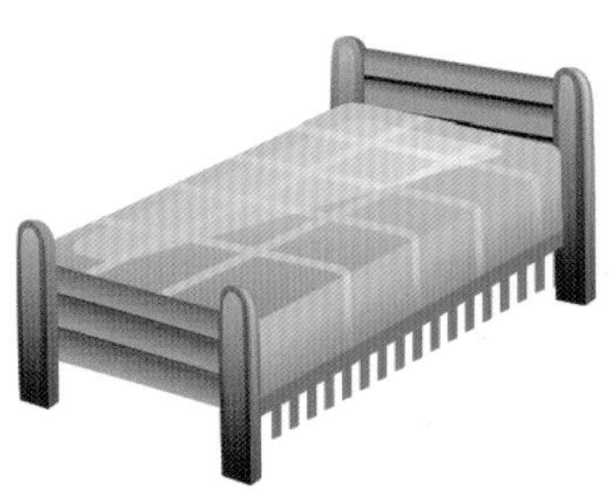

我是床。
(I'm a bed.)

我是水。
(I'm water.)

在冬天，你会穿着我。
我是什么？
（You wear me in winter.
What am I?）

鲸鱼生活在我里面。
我是什么？
（Whales live in me. What am I?）

你可以用我来做运动。
我是什么？
（You can use me to do exercise.
What am I?）

你喝我，我没有颜色。
我是什么？
（You drink me. I haven't got color.
What am I?）

你可以阅读我。
我是什么？
（You can read me. What am I?）

你可以住在我里面，
我有许多房间。我是什么？
（You can live in me, and I've got rooms. What am I?）

你在夜晚看不见我。
我是什么？
（You can't see me at night. What am I?）

你可以睡在我上面。
我是什么？
（You can sleep on me. What am I?）

我是太阳。
(I'm the sun.)

我是房子。
(I'm a house.)

我是网球。
(I'm tennis.)

我是海洋。
(I'm the sea.)

图文连线 — 我的房子

在每种生物与匹配的住处之间用直线连接起来。（Match each living being with their houses through a line.）

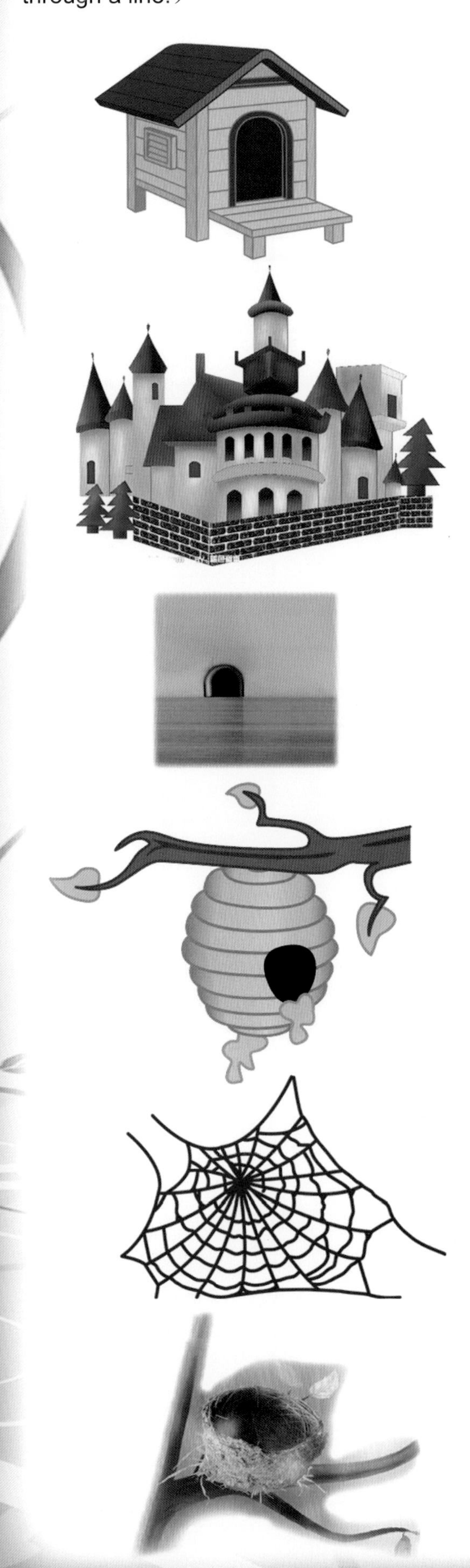

填充格子 — 红心

识别下面的坐标，并根据下面给出的颜色方案来涂色。（Identify the below coordinates and color the squares with the given colors.）

●涂红色的格子（Color these squares red）：2D、2E、2F、3C、3G、4C、4H、5D、5I、6C、6H、7C、7G、8D、8E、8F。

●涂粉红色的格子（Color these squares pink）：3D、3E、3F、4D、4E、4F、4G、5E、5F、5G、5H、6D、6E、6F、6G、7D、7E、7F。

	1	2	3	4	5	6	7	8	9
A									
B									
C									
D									
E									
F									
G									
H									
I									
J									
K									

电话聊天 — 简单对话

发出电话铃声。孩子必须拿起假想的话筒接电话。（Make a sound of a telephone ringing. The child must pick up an imaginary phone to answer your call.）

家长说左侧的话。孩子必须根据问题回答。（Say the questions from the left column. The child has to answer.）

最后，让孩子用铅笔连接下面图片的问题与答案。（At last, the child connects the questions and answers with a pencil.）

找出影子——三口之家

选出与方框中的图片精确匹配的影子。（Choose the shadow that matches the picture in the square exactly.）

各行各业——猜猜每人的职业

用下列职业名词来猜猜下表中各种人物的职业。（Guess the right job for each of these people with the following words.）

服务员（waiter）；机修工（mechanic）；消防员（fireman）；医生（doctor）；厨师（chef）；警察（policeman）

我维持秩序，并保护居民安全。
（I keep order and protect the citizens.）

____

我从火灾中救助人们和建筑物。
（I rescue people and buildings from fire.）

____

当有人生病的时候，
我让他们恢复健康。
（When people get sick,
I help them get well.）

____

我为你准备可口的食物。
（I prepare delicious food for you to eat.）

____

我负责点菜，并将食物和饮料端到桌子上。
（I take orders and serve food and drinks at table.）

____

我修理坏掉的车辆。
（I repair vehicles when they are damaged.）

____

发现物体——礼物

使用下面的密钥，给数字所在区域涂色，并发现图片中隐藏的物体。（According to the key below, color the areas by number and find the hidden object in the picture.）

22= 红色（red）；45= 棕色（brown）；15= 绿色（green）

过目不忘 — 辛勤的小园丁

仔细观察下面的图片一分钟，然后翻到下一页回答问题。（Look at the picture carefully for one minute, then turn to the next page to answer some questions.）

前一页的图片内容你能记得多少？不许偷看哦！
（How many things can you remember from the picture on the previous page? No peeking!）

1. 在门口的地上有什么东西？
（What is on the ground in front of the door?）

2. 女孩围着围巾吗？
（Is the girl wearing a scarf?）

3. 女孩的脚上穿的是什么？
（What does the girl have on her feet?）

4. 地上有玩具熊吗？
（Is there a toy bear on the ground?）

5. 在草地上有多少个花盆？
（How many flower pots are there on the lawn?）

6. 在凳子上有多少个花盆？
（How many flower pots are there on the stool ？）

7. 有花儿吗？
（Are there any flowers?）

8. 屋顶上有什么动物？
（What animal is on the roof?）

9. 洒水壶是什么颜色的？
（What color is the watering pot?）

找出不同——小朋友们在一起

找出下面两幅图片的八个不同之处。（Find the eight differences in the two pictures below.）

沿着线走—放风筝

马、熊、猪、企鹅各有什么样的风筝？（What kind of kite does each animal have?）

标注名称——身体部位

用下面的单词来标注人体各个部位。（Label the human body diagram using the words below.）

- 肩膀（shoulder）
- 胳膊（arm）
- 肘部（elbow）
- 手腕（wrist）
- 手（hand）
- 臀部（hip）
- 腿（leg）
- 脚后跟（heel）
- 脚（foot）
- 脑袋（head）
- 手指（fingers）
- 膝盖（knee）
- 脚趾（toes）
- 头发（hair）
- 脸颊（cheek）
- 眉毛（eyebrow）
- 前额（forehead）

拼接图片——梦幻家园

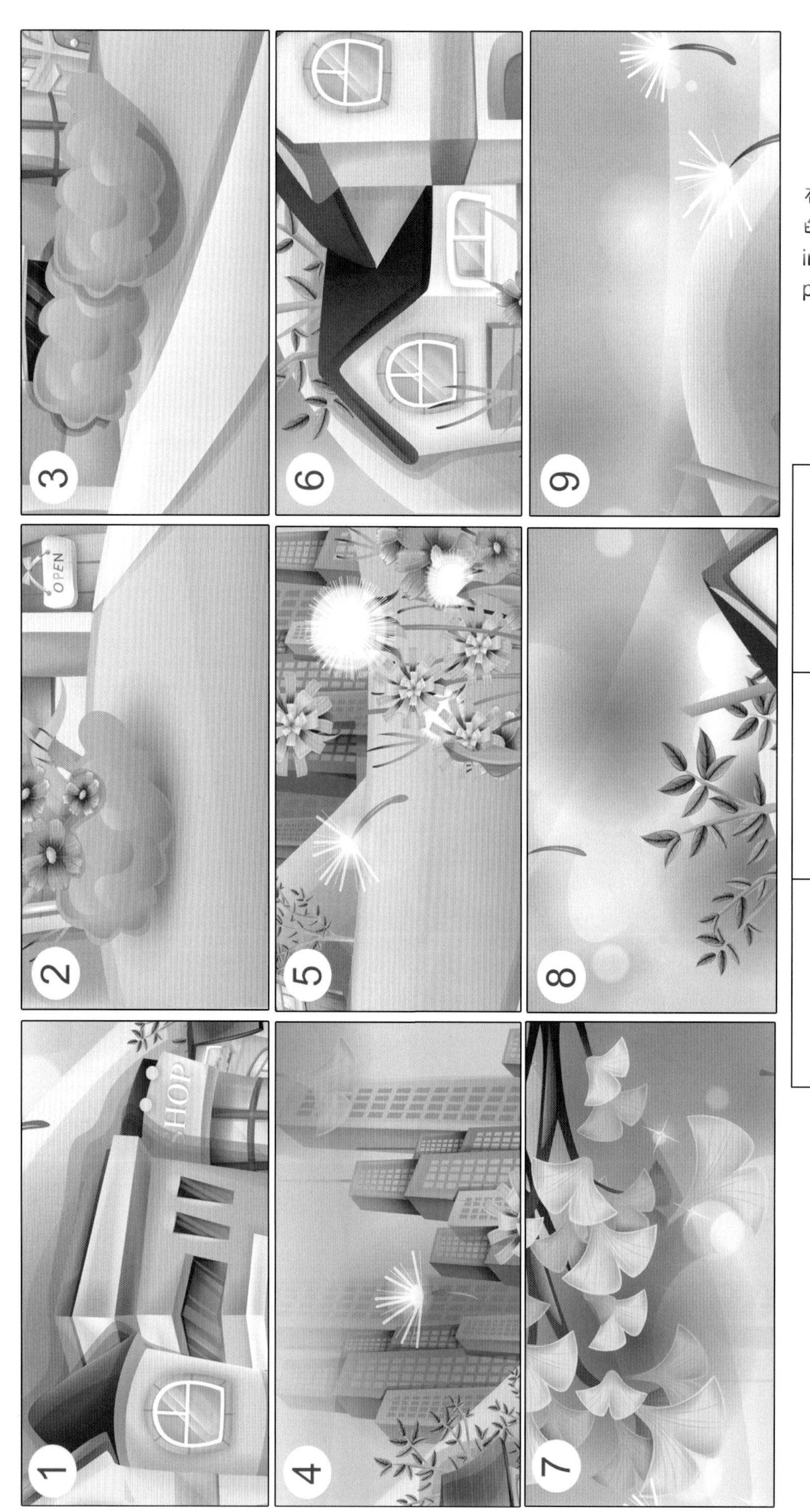

将左边图片的序号填写在空白处，拼接成一幅完整的图像。（Fill in the numbers in the blank to make a complete picture.）

		5

出入迷宫—找零食

哪个孩子能找到零食？（Which child can find the snack?）

临摹格子 — 蝙蝠侠

将下边网格中的图片临摹到上边的网格中。（Copy the bottom panels into the top panels, square by square.）

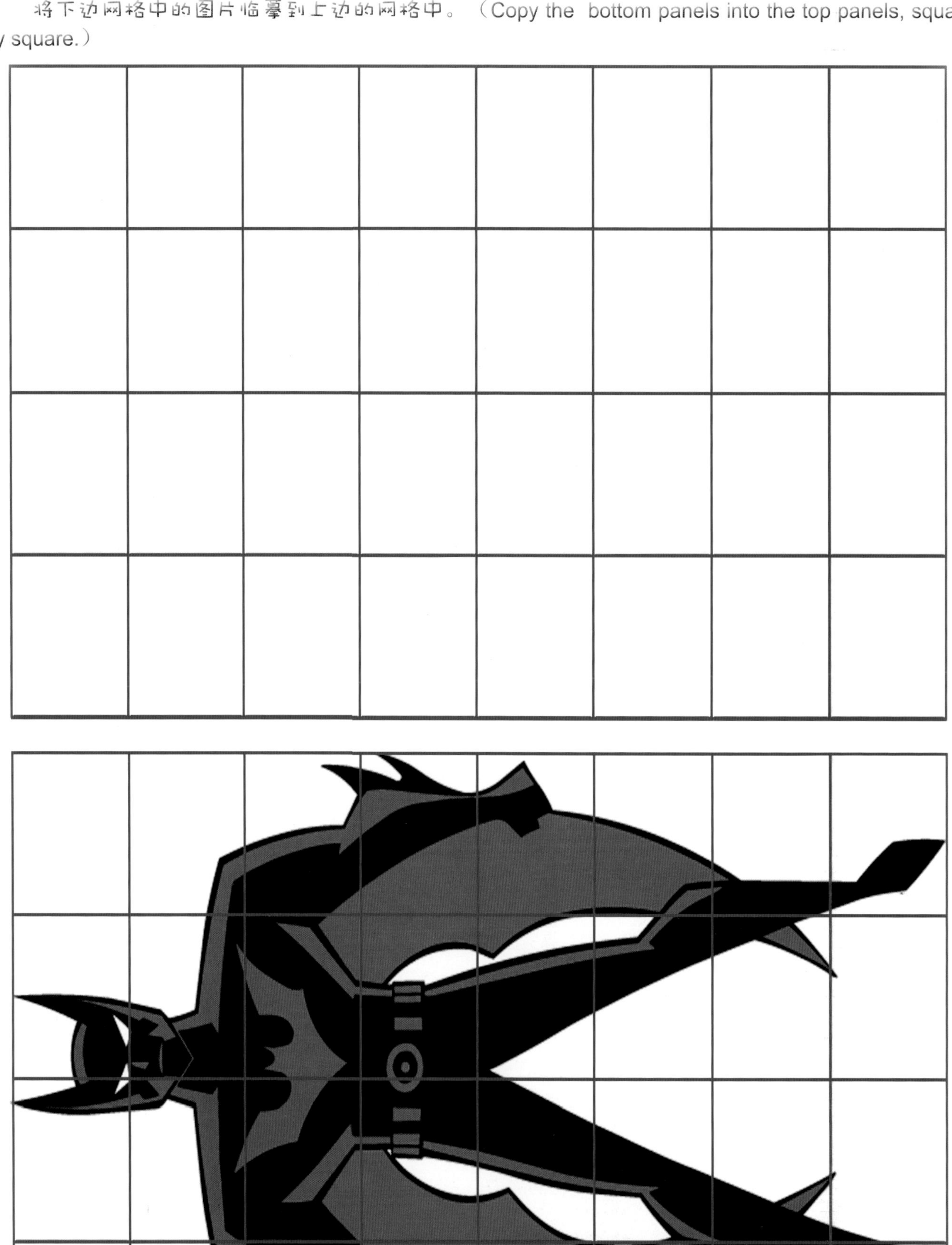

找出不同——乘车的小伙伴

找出下面两幅图片的六个不同之处。（Find the six differences in the two pictures below.）

色彩缤纷——客厅

请按照下图的颜色，给下页空白的客厅涂色。（Color the blank picture on the next page according to the colorful picture.）

12
9
3
6

找出相同——孩子们

在下面每一行中找出两幅相同的图片。（Find out the same two pictures in each line.）

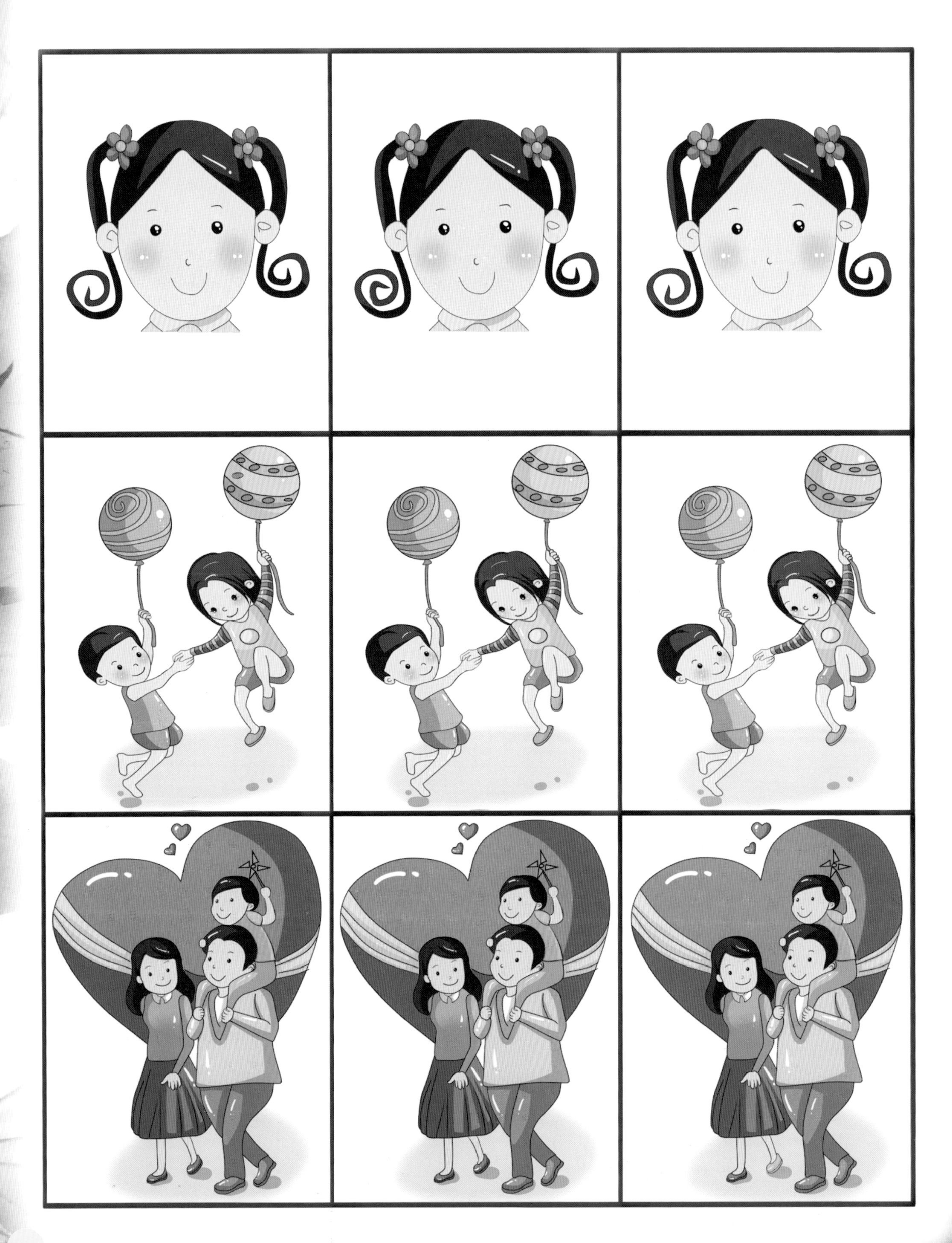

推断物品——找出交通工具

读下面的线索，推断哪个交通工具是正确的。在每个不符合线索的物体上画叉。然后圈选出正确的物体。（Read the clues below to figure out which vehicle is the correct one. Draw a X on each object that doesn't fit . Then circle the correct object.）

线索 1：此交通工具没有警报器。
（Clue 1: The vehicle does not have an alarm.）

线索 2：此交通工具没有螺旋桨。
（Clue 2: The vehicle does not have a propeller.）

线索 3：此交通工具不用于建筑。
（Clue 3: The vehicle is not used for construction.）

线索 4：此交通工具不用于清洁道路。
（Clue 4: The vehicle is not used for cleaning.）

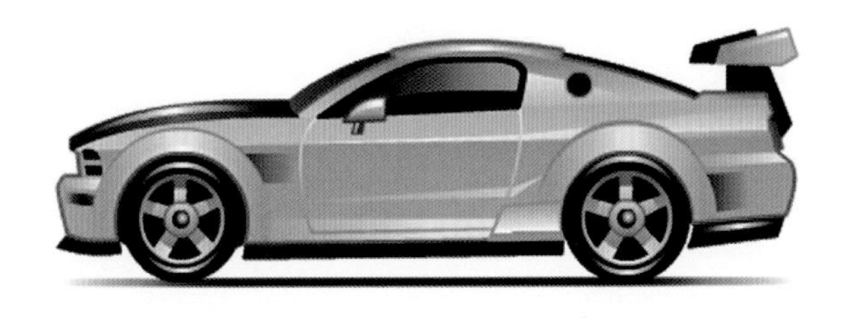

看图猜谜—生活谜语

根据下面带有插图的答案来猜谜语。（Find the answers to the riddles.）

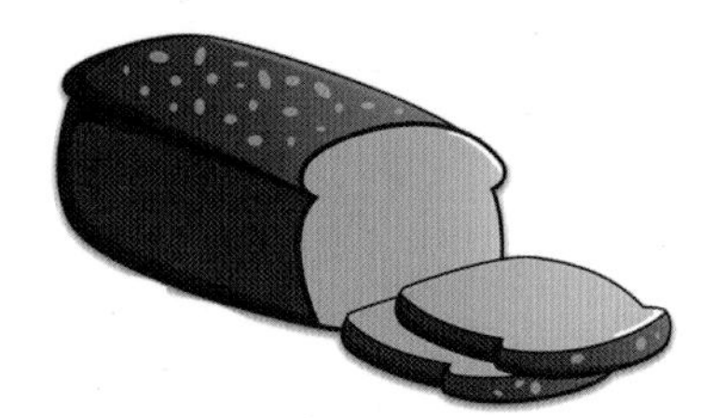

你可以喝我。我是白色的。
我是什么？
（You can drink me. I'm white. What am I?）

我是面包。
（I'm bread.）

你可以骑我。我是什么？
（You can ride me. What am I?）

我是椅子。
（I'm a chair.）

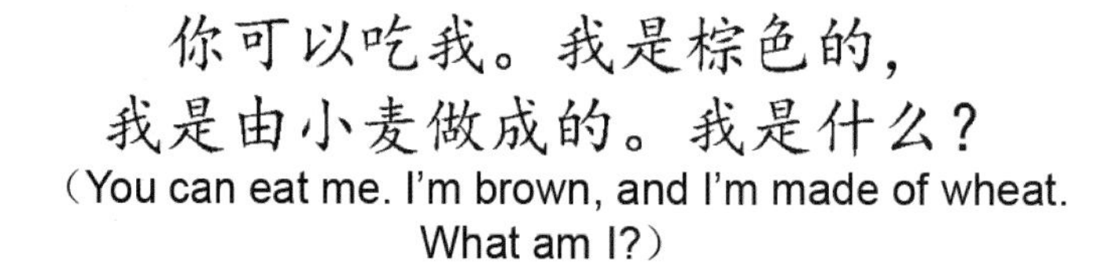

你可以吃我。我是棕色的，
我是由小麦做成的。我是什么？
（You can eat me. I'm brown, and I'm made of wheat.
What am I?）

你可以画出我。我是什么？
（You can draw or paint me. What am I?）

我是牛奶。
（I'm milk.）

我是头发。
（I'm hair.）

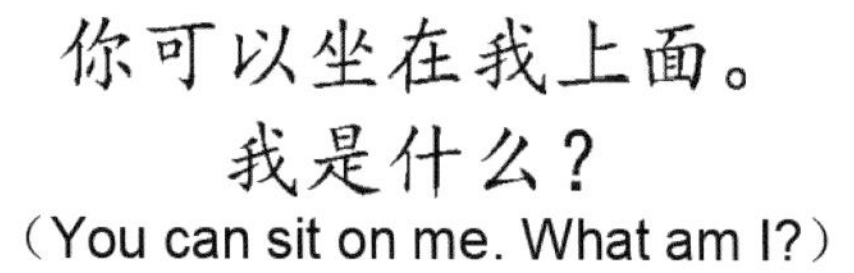

你可以坐在我上面。
我是什么？
（You can sit on me. What am I?）

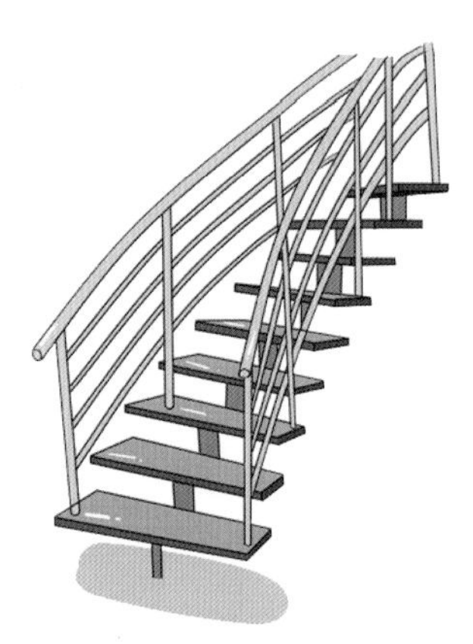

你可以把我戴在你的头上。
我是什么？
（You can wear me on your head. What am I?）

我是自行车。
（I'm a bike.）

我是楼梯。
（I'm stairs.）

你可以在房子里看见我，
你在我上面走上走下的。
我是什么？
（You can find me in a house, and
you go up and down me. What am I?）

我是帽子。
（I'm a hat.）

我就长在你的头上。
我是什么？
（You have got me on your head. What am I?）

我是图画。
（I'm a picture.）

图文连线——我住的地方

在每种生物与匹配的住处之间画一条线。（Draw a line to match each living being with their house.）

熊生活在____中。（The bear lives in a _____.）

海洋（sea）

松鼠生活上在____上。（The squirrel lives in a _____.）

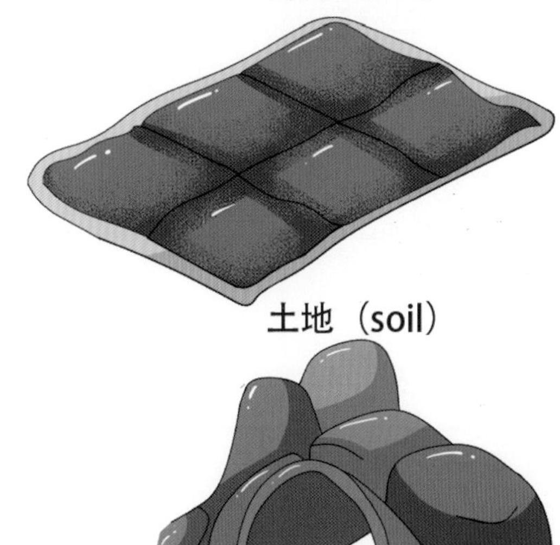

土地（soil）

鲸鱼生活在____中。（The whale lives in the _____.）

洞穴（cave）

蚂蚁生活在____里。（The ant lives in the _____.）

房子（house）

我生活在____里。（I live in a _____.）

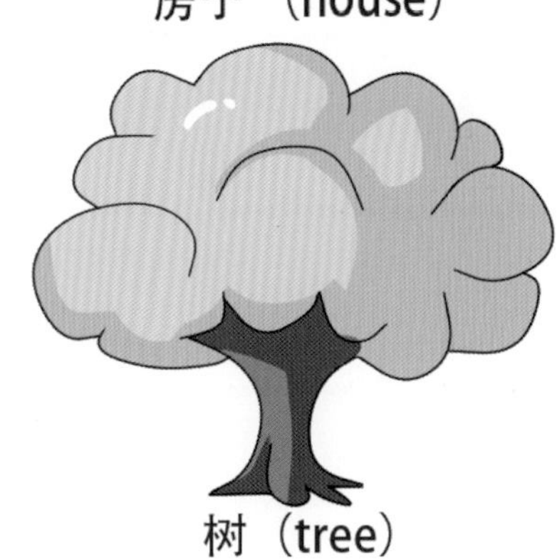

树（tree）

填充格子一房子

识别下面的坐标，并根据下面给出的颜色方案来涂色。（Identify the below coordinates and color the squares with the given colors.）

●涂黑色的格子（Color these squares black）：1E、2D、3C、4B、5A、6B、7C、8D、9E。

●涂红色的格子（Color these squares red）：2E、2F、2G、2H、2I、2J、3D、3E、3F、3G、3H、3I、3J、4C、4F、4G、4H、4I、4J、5B、5C、5F、5G、6C、6F、6G、6H、6I、6J、7D、7E、7F、7G、7H、7I、7J、8E、8F、8G、8H、8I、8J。

●涂黄色的格子（Color these squares yellow）：4D、4E、5D、5E、6D、6E。

●涂蓝色的格子（Color these squares blue）：5H、5I、5J、6H、6I、6J。

●涂绿色的格子（Color these squares green）：1K、2K、3K、4K、5K、6K、7K、8K、9K。

	1	2	3	4	5	6	7	8	9
A									
B									
C									
D									
E									
F									
G									
H									
I									
J									
K									

电话聊天 — 日常对话

发出电话铃声。孩子必须拿起假想的话筒接电话。（Make a sound of a telephone ringing. The child must pick up an imaginary phone to answer your call.）

家长说左侧的话。孩子必须根据问题回答。（Say the questions from the left column. The child has to answer.）

最后，让孩子用铅笔连接下面图片的问题与答案。（At last, the child connects the questions and answers with a pencil.）

找出影子 — 玩跷跷板

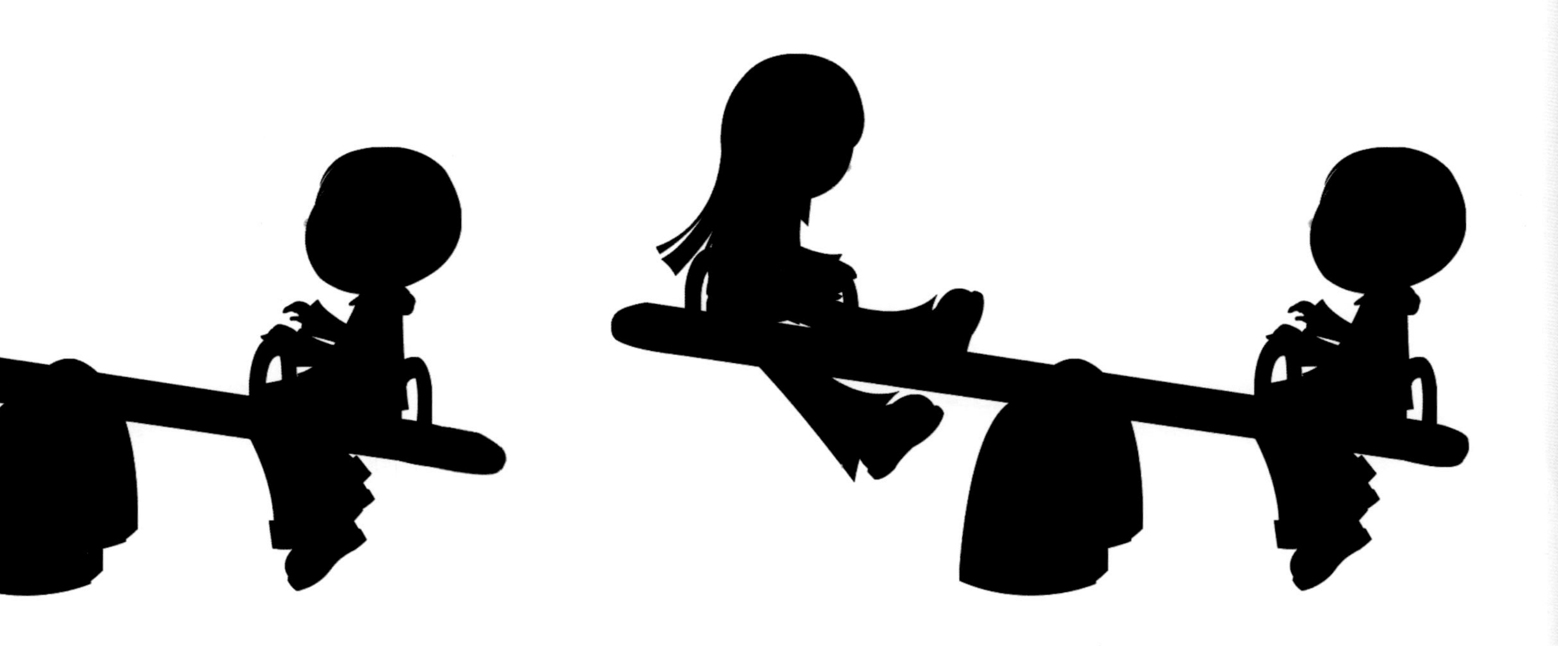

选出与方框中的图片精确匹配的影子。（Choose the shadow that matches the picture in the square exactly.）

各行各业 — 猜猜每人的职业

用下列职业名词来猜猜下表中各种人物的职业。（Guess the right job for each of these people with the following words.）

运动员（athlete）；艺术家（artist）；农民（farmer）；飞行员（pilot）；摄影师（photographer）；音乐家（musician）

我种植庄稼，并饲养家畜。
（I grow crops and raise livestock.）

______

我给人物、场地或物品拍照。
（I take pictures of people, places, or things.）

______

我创作并演奏音乐。
((I create and perform music.）

______

我驾驶飞机运送乘客。
（I fly an airplane to carry passengers.）

______

我在体育活动中取得好成绩。
（ I try to get good results in sport.）

______

我创作艺术品。
（I create art works.）

______

发现物体——城堡

使用下面的密钥，给数字所在区域涂色，并发现图片中隐藏的物体。（According to the key below, color the areas by number and find the hidden object in the picture.）

27= 蓝色（blue）；21= 红色（red）；77= 黄色（yellow）；58= 紫色（purple）；89= 黑色（black）；44= 橙色（orange）

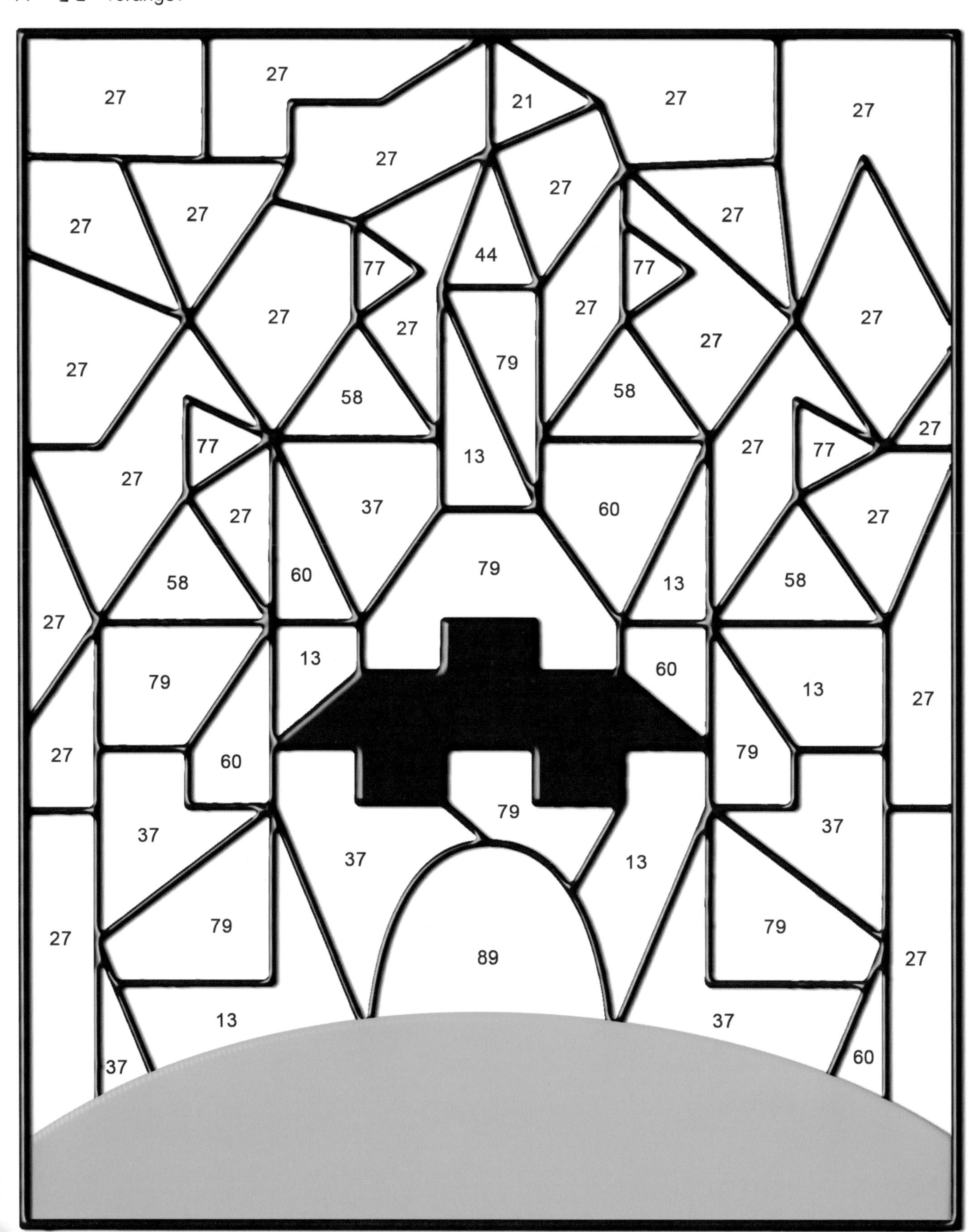

标注名称一家庭

在下图中标注家庭成员，如女儿、儿子、爸爸、妈妈、爷爷、奶奶。（Label the following family members in the picture: daughter, son, father, mother, grandfather, grandmother.）

拼接图片 — 幸福一家人

将右边图片的序号填写在空白处，拼接成一幅完整的图像。（Fill in the numbers in the blank to make a complete picture.）

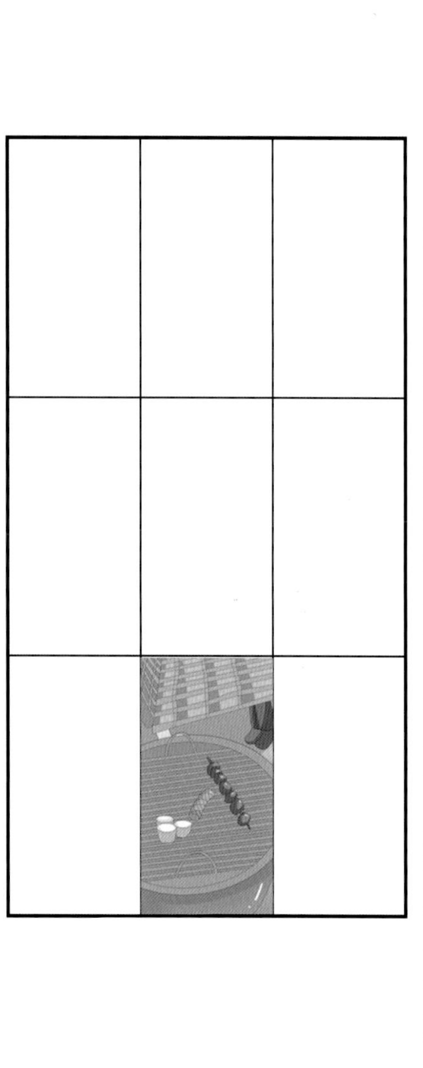

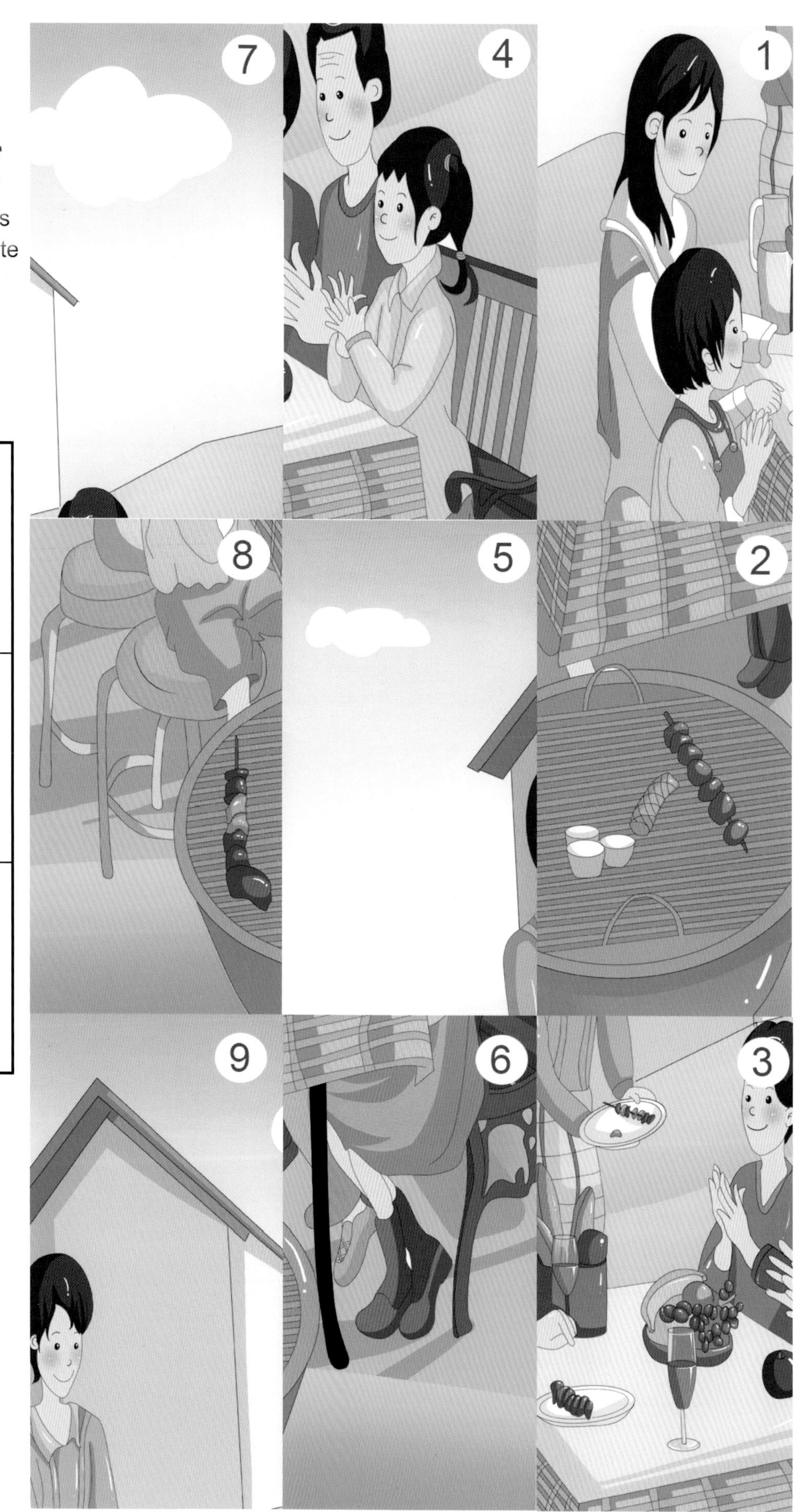

色彩缤纷—上班族

随意给下图人物涂色。（Color the picture below as you wish.）

纵横字谜——身体部位

用给出的线索，完成身体部位的纵横字谜。（Complete the body parts crossword using the given clues.）

	1		2				
3							
			4				
		5					
					9		
6							
		7		8			
	10						

横向：

1. arm（胳膊）
3. foot（脚）
4. hair（头发）
6. eyes（眼睛）
7. hand（手）
10. ears（耳朵）

纵向：

2. mouth（嘴巴）
3. finger（手指）
5. teeth（牙齿）
8. nose（鼻子）
9. head（脑袋）

出入迷宫—警察

画出从起点到终点的路径。（Draw a line from the start to the end.）

推断物品——选择交通工具

请读下面的小故事，然后从图片中圈选出正确的物体。（Please read the short story below and then circle the correct object from the pictures.）

一个男人要去山顶上的某个建筑。但是没有去山顶的路。在山顶上的建筑旁边有一个停机坪。那么这个男人应该用什么交通工具去那里呢？（A man needs to get to a building at the top of a mountain. There are no roads that lead to the top of the mountain. There is a landing pad on top of the mountain next to the building. What vehicle should the man use to get there?）

亲子手工——三维立体房子

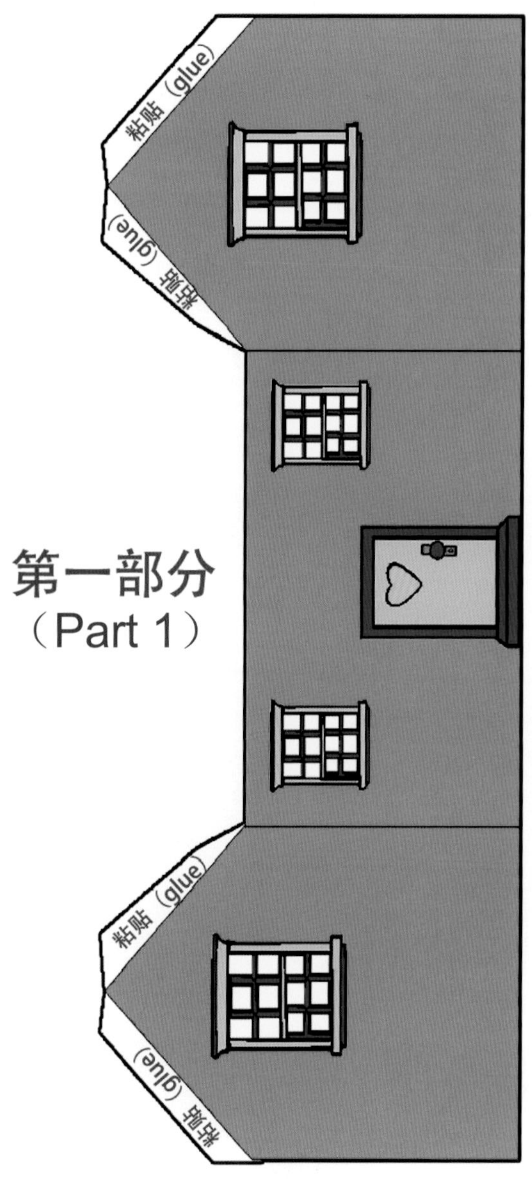

第一部分
（Part 1）

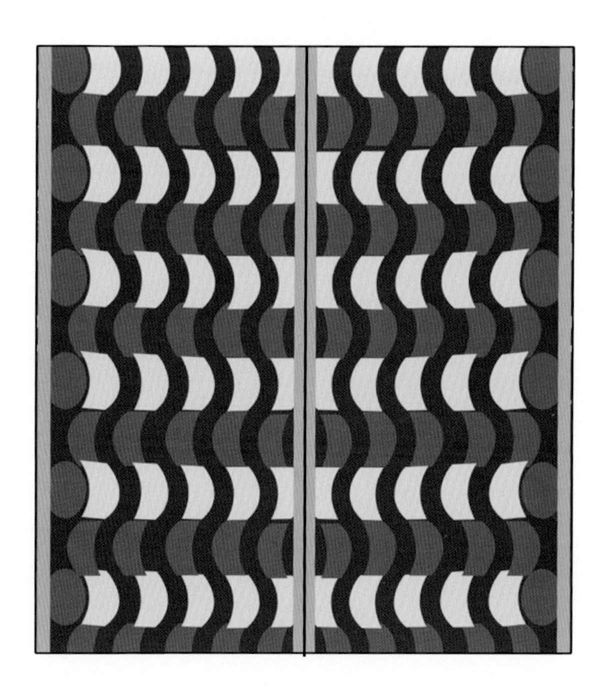

第三部分
（Part 3）

按照下面步骤制作三维立体房子（Make a 3D house craft according to the steps below）：

●剪下这三部分。（Cut out the three parts.）

●沿着所有实线折叠。（Fold along all of the solid lines.）

●按照图中指示进行粘贴。（Glue where indicated on the tabs.）

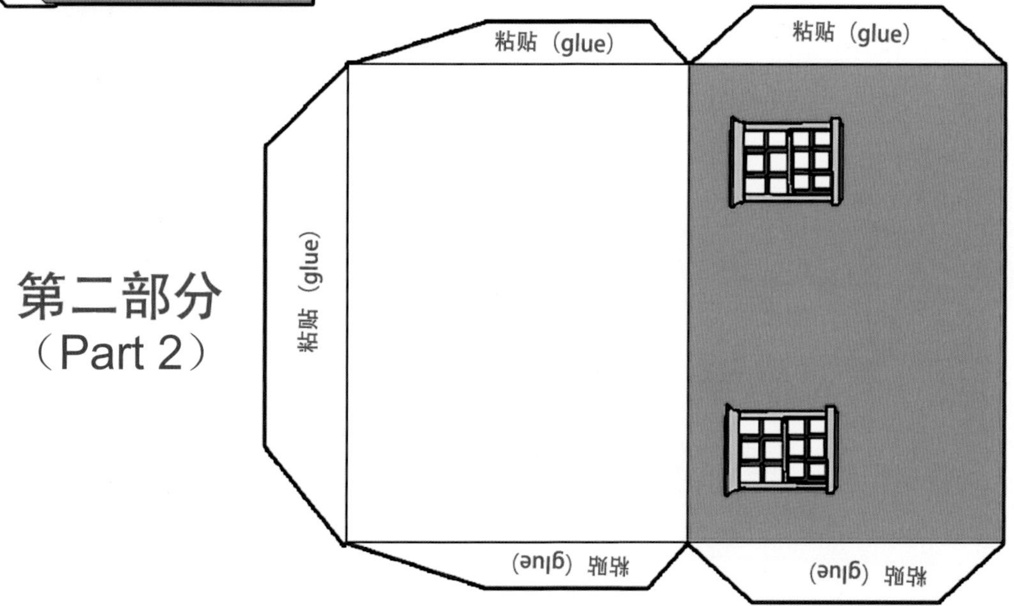

第二部分
（Part 2）

亲子手工

三维立体房子

（Let’s make a 3D house craft.）

对称画画—坐着的儿童

利用对称的格子画完儿童图像。（Finish off the picture of the child using the grid to help you.）

看图故事 — 繁忙的蜘蛛

家长给孩子讲下面的故事，同时给孩子指出下页图中的各种动物。（Tell the following story to the child, and point out the relative animals on the next page.）

一天早晨，田野上的风吹来了一只蜘蛛，她拖着一条细细的、柔滑的丝。（Early one morning the wind blew a spider across the field, a thin, silky thread trailed from her body.）

蜘蛛落在一个靠近农场院子的栅栏柱上，并开始用她的柔滑的丝结网。（The spider landed on a fencepost near a farm yard and began to spin a web with her silky thread.）

马儿说："想去兜兜风吗？"（"Neigh! Neigh!" said the horse, "Want to go for a ride?"）
蜘蛛没有回答，她还忙着结网。（The spider didn't answer. She was very busy spinning her web.）

奶牛哞哞地说："想吃些草吗？"（"Moo! Moo!" said the cow, "Want to eat some grass?"）
蜘蛛没有回答，她还忙着结网。（The spider didn't answer. She was very busy spinning her web.）

绵羊咩咩地说："想在草地上跑吗？"（"Baa! Baa!" said the sheep, "Want to run in the meadow?"）
蜘蛛没有回答，她还忙着结网。（The spider didn't answer. She was very busy spinning her web.）

山羊叫着说："想在石头上跳吗？"（"Maa! Maa!" said the goat, "Want to jump on the rocks?"）
蜘蛛没有回答，她还忙着结网。（The spider didn't answer. She was very busy spinning her web.）

猪呼噜地说："想在泥巴里打滚吗？"（"Oink! Oink!" grunted the pig, "Want to roll in the mud?"）
蜘蛛没有回答，她还忙着结网。（The spider didn't answer. She was very busy spinning her web.）

狗汪汪叫着说："想去追猫吗？"（"Woof! Woof!" barked the dog, "Want to chase a cat?"）
蜘蛛没有回答，她还忙着结网。（The spider didn't answer. She was very busy spinning her web.）

猫喵喵地说："想去打个盹吗？"（"Meow! Meow!" said the cat, "Want to take a nap?"）
蜘蛛没有回答，她还忙着结网。（The spider didn't answer. She was very busy spinning her web.）

鸭子呱呱叫着说："想去游泳吗？"（"Quack! Quack!" called the duck, "Want to go for a swim?"）
蜘蛛没有回答，她还忙着结网。（The spider didn't answer. She was very busy spinning her web.）

公鸡啼叫着说："想去抓一只苍蝇吗？"（"Cock-a-doodle do!" crowed the rooster, "Want to catch a fly?"）
蜘蛛没有回答，她还忙着结网。（The spider didn't answer. She was very busy spinning her web.）

然后蜘蛛在她的网里抓住了苍蝇……就是那样！（And the spider caught the fly in her web ... just like that!）

猫头鹰呜呜地说："谁建了这个漂亮的网？"（"Whoo? Whoo?" asked the owl, "Who built this beautiful web?"）
蜘蛛没有回答。她已经睡着了。（The spider didn't answer. She had fallen asleep.）
这真是非常非常忙的一天。（It had been a very, very busy day.）

出入迷宫——找到篮球与足球

帮助篮球运动员找到篮球，再帮助足球运动员找到足球。（Help the basketball player find the basketball, and then help the football player find the football.）

填充图片——打乒乓球

将含有两个点的区域涂成黄色；将含有一个点的区域涂成蓝色；将含有一个方块的区域涂成灰色。你发现了什么隐藏的图形？（Color the areas that have two dots yellow. Color the areas that have one dot blue. Color the areas that have one square gray. What hidden picture can you find?）

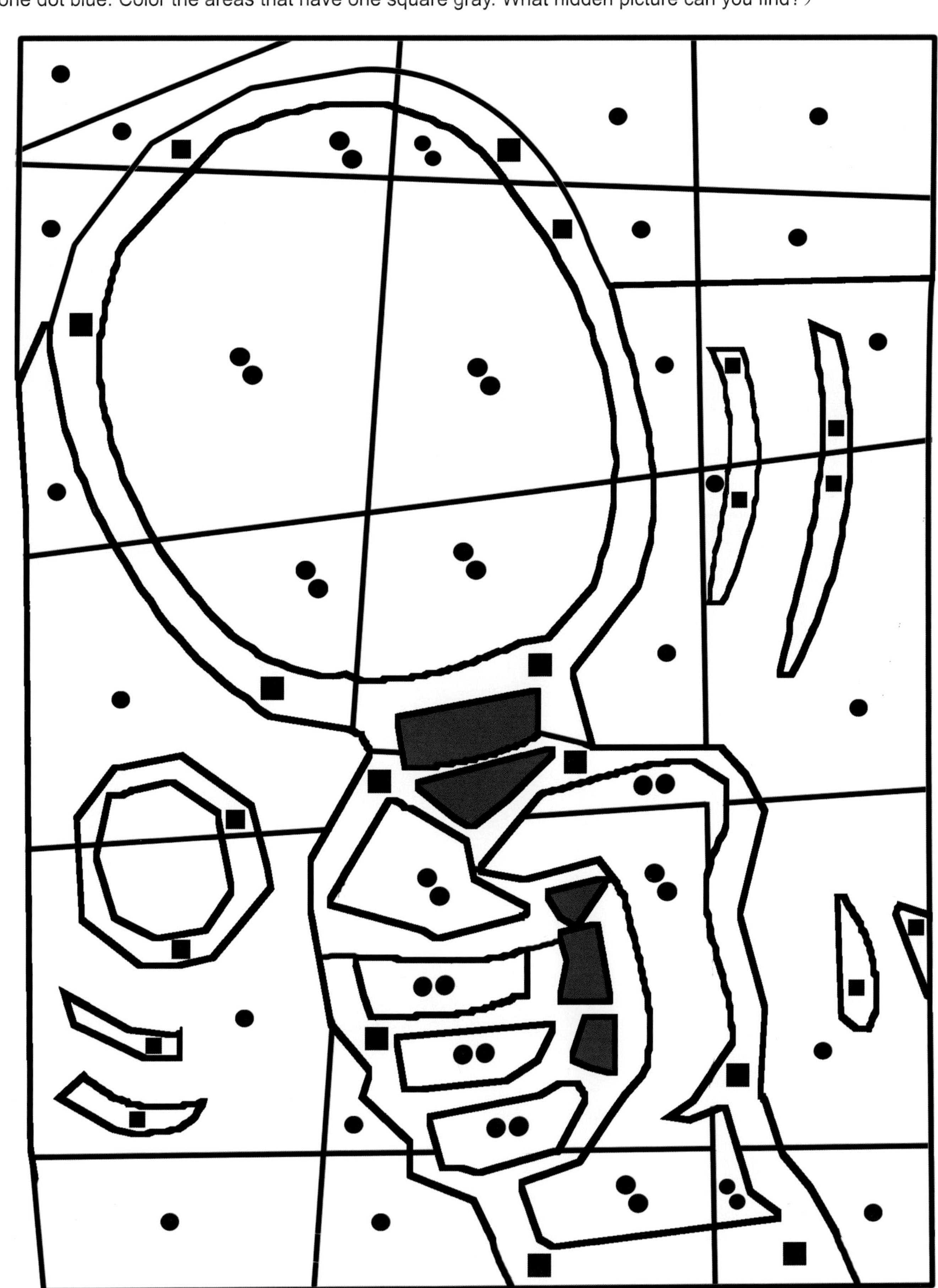

沿着线走 — 谁能钓到鱼

沿着线看看谁能钓到这条鱼。（Follow the trails to see who will get the fish.）

找出相同—扑克牌

你手上抓着一把扑克 K，其中有两张是完全相同的。相同的是哪两张呢？（You are holding a handful of Kings, and two of them are exactly the same. Which two are they?）

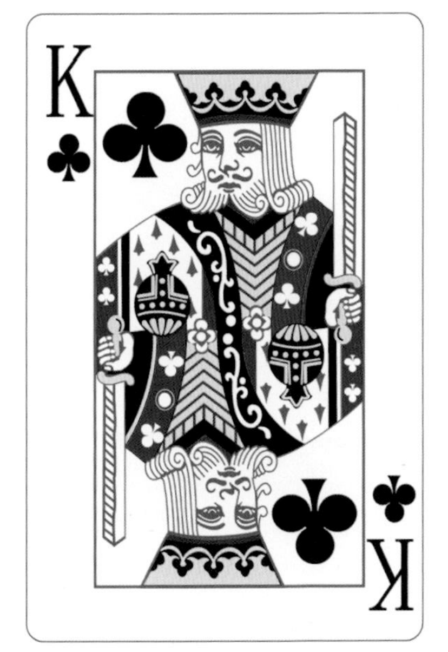

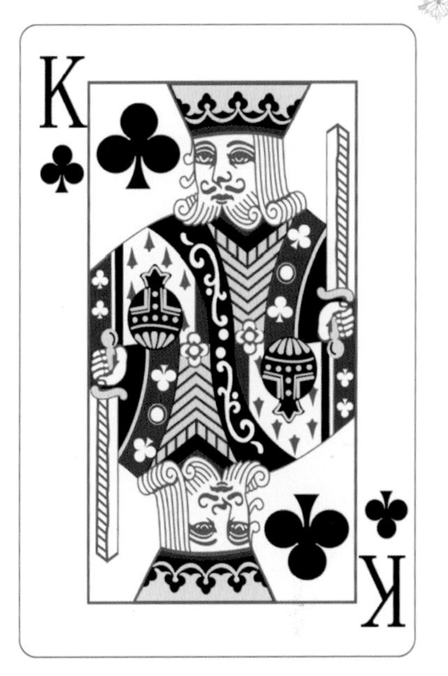

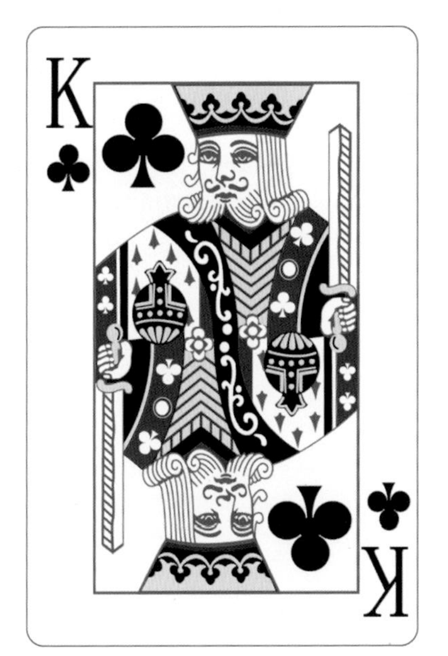

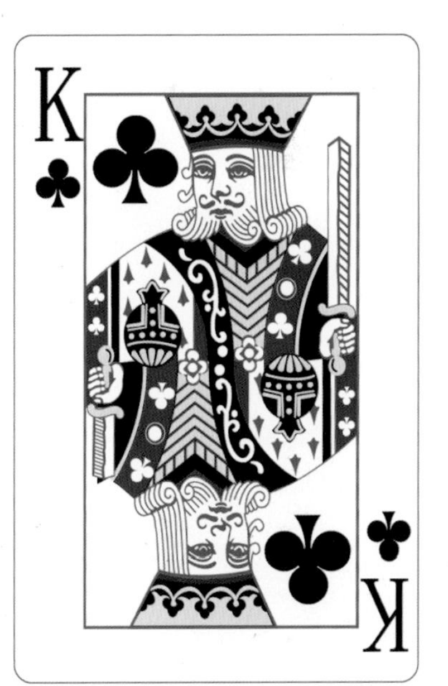

匹配连线 — 学校物品

将单词与图片用直线连接起来。（Match the words to their pictures through a line.）

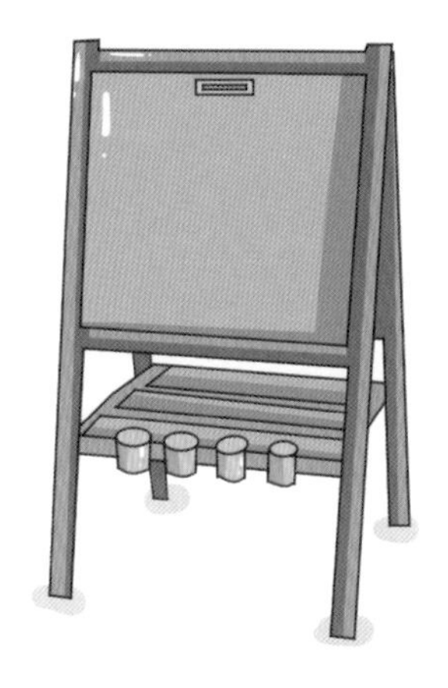

椅子
(chair)

笔记本
(notebook)

老师
(teacher)

画架
(easel)

画笔
(brush)

黑板
(blackboard)

校车
(schoolbus)

橡皮擦
(eraser)

卷笔刀
(sharpener)

地球仪
(globe)

书架
(bookshelf)

曲别针
(paperclip)

粉笔
(chalks)

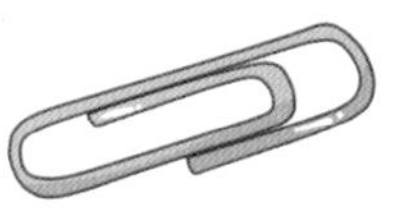

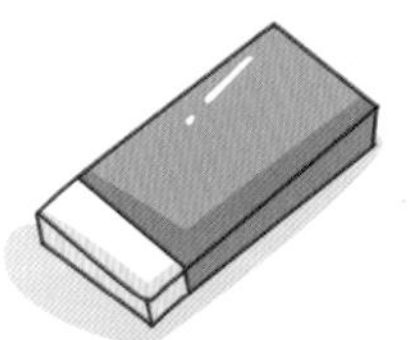

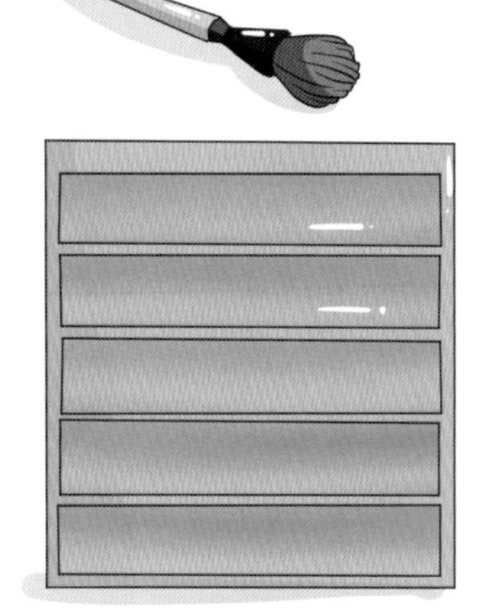

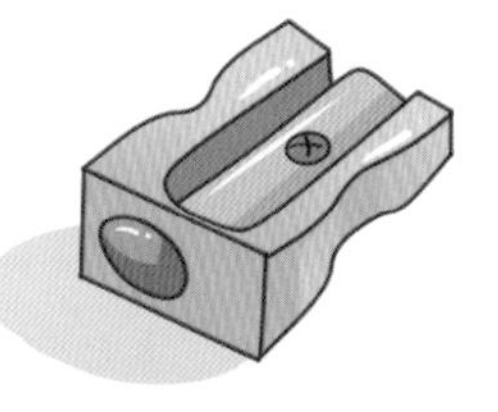

答案与提示

P1

P2

1. 不是（No）。 2. 不是（No）。
3. 是（Yes）。 4. 不是（No）。
5. 不是（No）。 6. 是（Yes）。
7. 不是（No）。 8. 是（Yes）。
9. 不是（No）。 10. 是（Yes）。
11. 不是（No）。 12. 不是（No）。
13. 是（Yes）。 14. 不是（No）。
15. 是（Yes）。

P3

P5

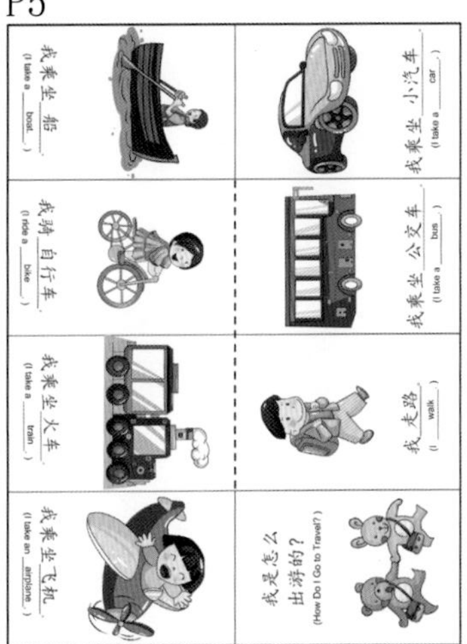

P7

P8

P9

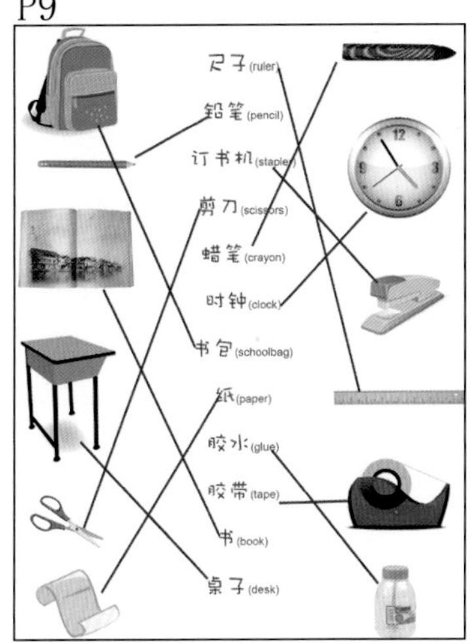

P14

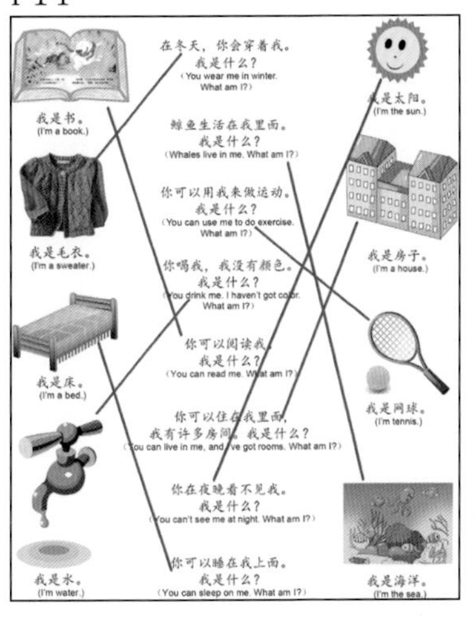

P15

P16

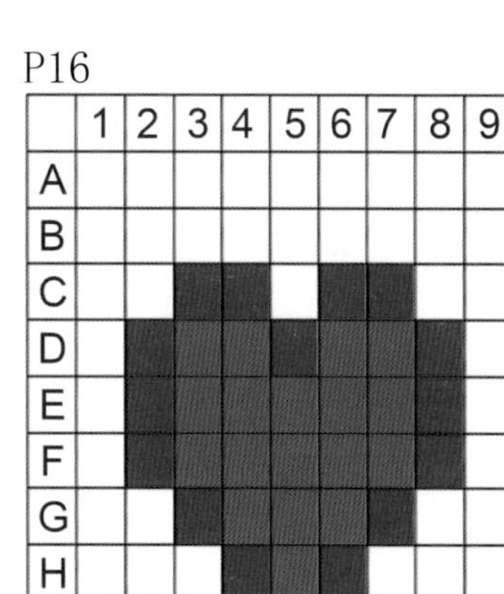

P17

P18 有红色对号的是完全匹配的影子。蓝色圈选之处是不同的地方。

P19

P20

P22

1. 一个足球（A football）
2. 没有（No）
3. 靴子（Boots）
4. 是的（Yes）
5. 9（Nine）
6. 5（Five）
7. 没有（No）
8. 一只红色鹦鹉（A red parrot）
9. 红色（Red）

P23

P24

1. 马有一个燕子风筝。（The horse has a swallow kite.）
2. 熊有一个蜻蜓风筝。（The bear has a dragonfly kite.）
3. 猪有一个金鱼风筝。（The pig has a goldfish kite.）
4. 企鹅有一个蝴蝶风筝。（The penguin has a butterfly kite.）

P24

P25

P26

P27

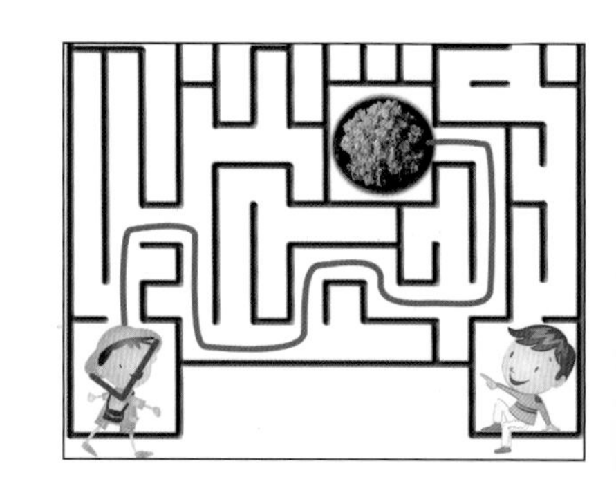

P28

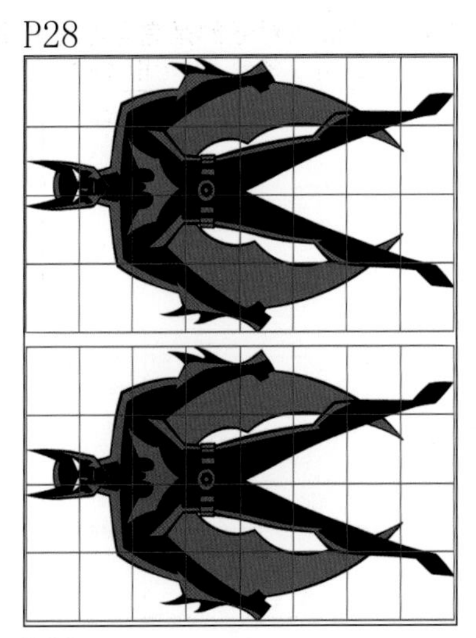

P29

P32

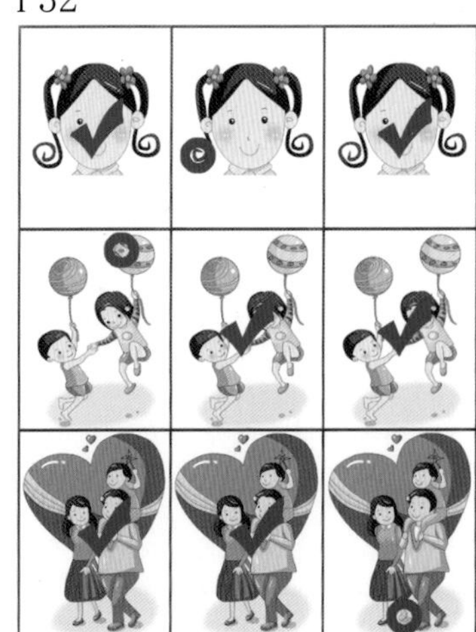

P33

P34

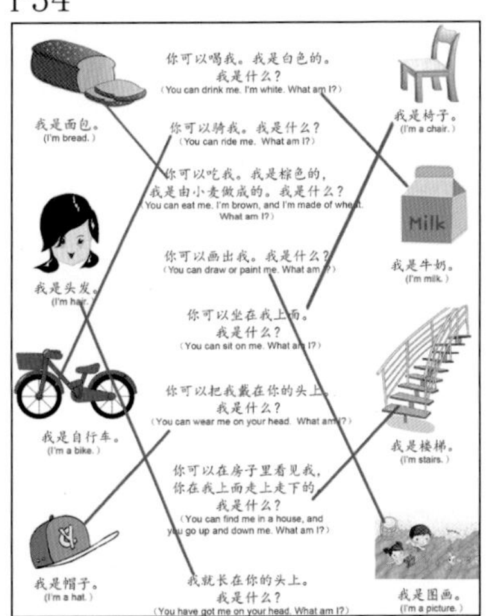

P35

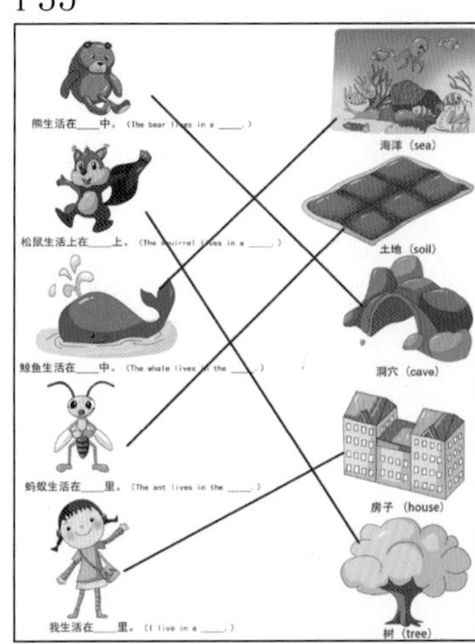

P36

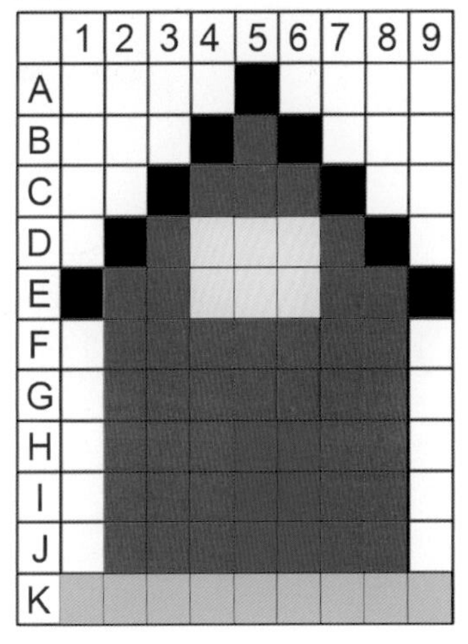

P37

P38 有红色对号的是完全匹配的影子。蓝色圈选之处是不匹配的地方。

P39

P40

P41

P42

P43

P44

P45

P46

P47

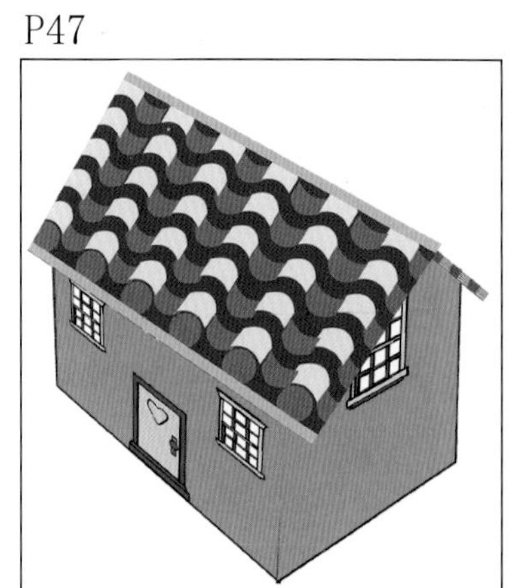

P49

P52

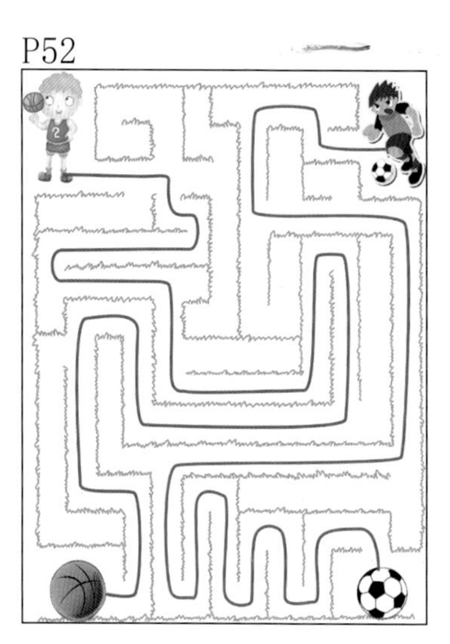

P53

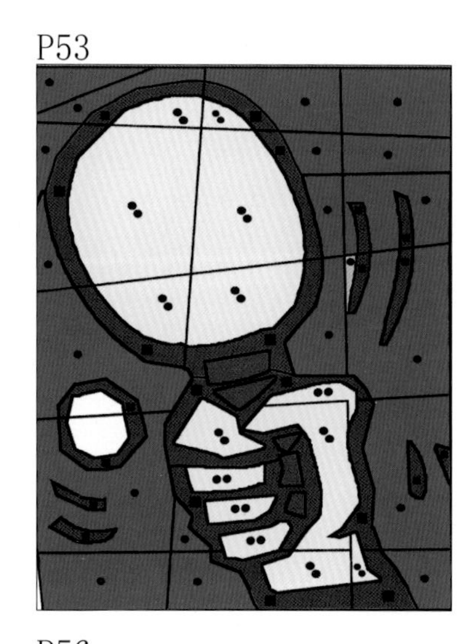

P54

P55 有红色对号的是完全相同图形。蓝色圈选之处是不同的地方。

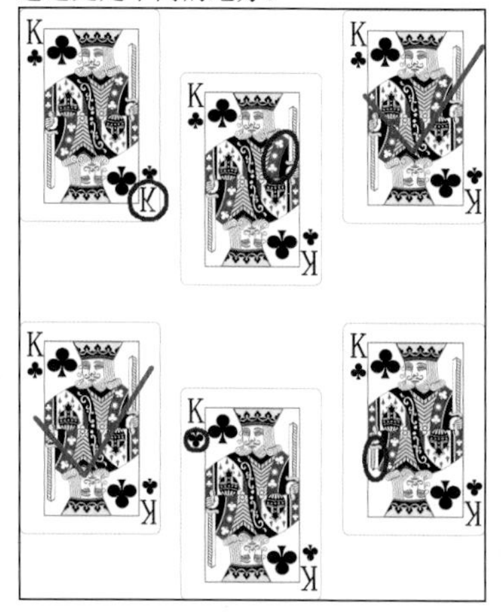

P56

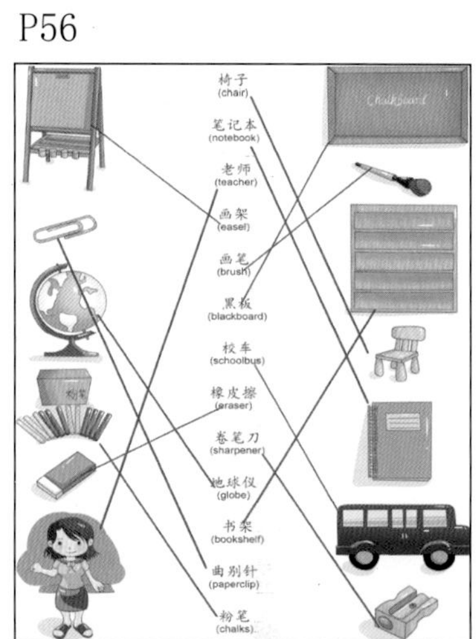